Neighborhood Safari

Opossums

by Martha London

www.focusreaders.com

Copyright © 2021 by Focus Readers®, Lake Elmo, MN 55042. All rights reserved. No part of this book may be reproduced or utilized in any form or by any means without written permission from the publisher.

Focus Readers is distributed by North Star Editions: sales@northstareditions.com | 888-417-0195

Produced for Focus Readers by Red Line Editorial.

Photographs ©: Shutterstock Images, cover, 1, 4, 7, 11, 12, 15, 17, 18; iStockphoto, 8, 21 (opossums); Red Line Editorial, 21 (chart)

Library of Congress Cataloging-in-Publication Data
Names: London, Martha, author.
Title: Opossums / by Martha London.
Description: Lake Elmo, MN : Focus Readers, [2021] | Series: Neighborhood safari | Includes index. | Audience: Grades 2-3
Identifiers: LCCN 2019060196 (print) | LCCN 2019060197 (ebook) | ISBN 9781644933534 (hardcover) | ISBN 9781644934296 (paperback) | ISBN 9781644935811 (pdf) | ISBN 9781644935057 (ebook)
Subjects: LCSH: Opossums--Juvenile literature.
Classification: LCC QL737.M34 L66 2021 (print) | LCC QL737.M34 (ebook) | DDC 599.2/3--dc23
LC record available at https://lccn.loc.gov/2019060196
LC ebook record available at https://lccn.loc.gov/2019060197

Printed in the United States of America
Mankato, MN
082020

About the Author

Martha London writes books for young readers. When she's not writing, you can find her hiking in the woods.

Table of Contents

CHAPTER 1
Trees and Fields 5

CHAPTER 2
Fur and Claws 9

CHAPTER 3
Finding Food 13

THAT'S AMAZING!
Playing Possum 16

CHAPTER 4
Behavior 19

Focus on Opossums • 22
Glossary • 23
To Learn More • 24
Index • 24

Date: 2/2/22

J 599.276 LON
London, Martha,
Opossums /

PALM BEACH COUNTY
LIBRARY SYSTEM
3650 SUMMIT BLVD.
WEST PALM BEACH, FL 33406

Chapter 1

Trees and Fields

An opossum walks slowly across a field. Then it climbs up a tree. It uses its long tail to grip the branches.

Opossums live in North America. They live near forests or fields. Many opossums live near water. They are good swimmers.

Some opossums live near people's houses. They can cause problems by digging in trash. If people see an opossum, they should leave it alone.

Chapter 2

Fur and Claws

Many opossums have gray or white fur. The fur covers their bodies. But they have no fur on their tails.

Opossums have long snouts. They have sharp claws. Their back feet have toes that are similar to thumbs. Opossums use their feet to grip trees as they climb. Their tails can wrap around branches, too.

Fun Fact: Some opossums can weigh 12 pounds (5.4 kg).

Chapter 3

Finding Food

Opossums usually come out at night. They use their sense of smell to find food. Opossums eat insects, seeds, and plants.

Some opossums are **scavengers**. They eat whatever they can find. They even eat dead animals. Sometimes they hunt. They catch mice and birds. Opossums use their sharp teeth to chew their food.

Fun Fact

An opossum has 50 teeth in its mouth.

That's Amazing!

Playing Possum

Opossums have short legs. They cannot run fast. If an opossum is in danger, its body goes **limp**. Its tongue hangs out of its mouth. The opossum even gives off a bad smell. These actions confuse **predators**. Predators usually leave the opossum alone.

Chapter 4

Behavior

Many opossums spend a lot of time in trees. They climb high to stay safe. Opossums can also have **dens** on the ground.

Opossums fill their dens with leaves. They use the dens to sleep and have babies.

Opossums are **marsupials**. Female opossums have pouches. They carry their babies inside. Baby opossums are called **joeys**.

Fun Fact: A newborn opossum is the size of a bee.

Life Cycle

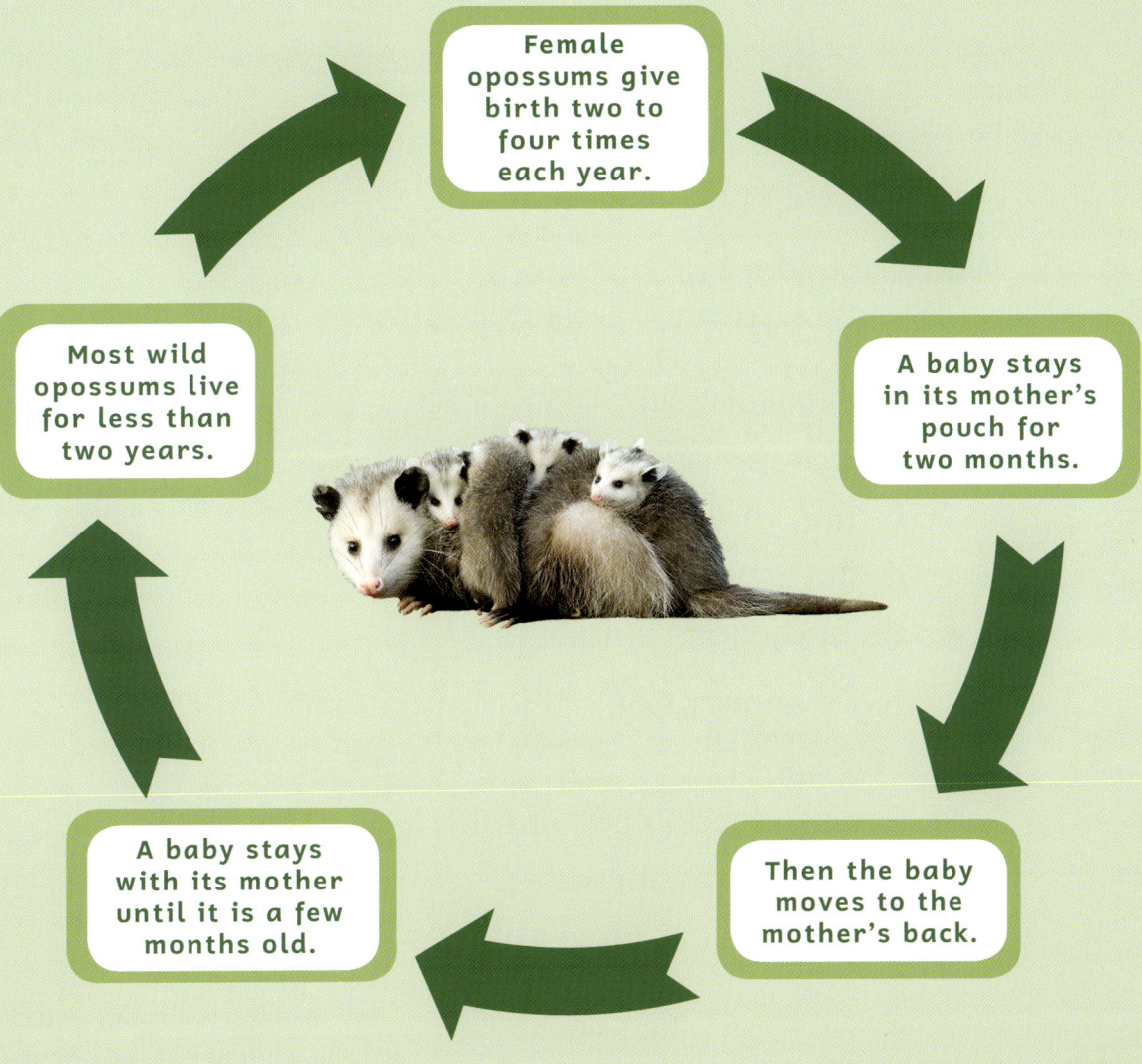

- Female opossums give birth two to four times each year.
- A baby stays in its mother's pouch for two months.
- Then the baby moves to the mother's back.
- A baby stays with its mother until it is a few months old.
- Most wild opossums live for less than two years.

FOCUS ON
Opossums

Write your answers on a separate piece of paper.

1. Write a few sentences describing what an opossum looks like.

2. Would you want an opossum to live in your neighborhood? Why or why not?

3. How does an opossum stay safe from predators?
   - A. It runs very fast.
   - B. It fights back.
   - C. Its body goes limp.

4. Why might some people think opossums are dirty?
   - A. Opossums don't have fur.
   - B. Opossums avoid water.
   - C. Opossums eat dead animals and trash.

Answer key on page 24.

Glossary

dens
The homes of wild animals.

joeys
Baby marsupials.

limp
Loose and floppy.

marsupials
Animals that raise their young in pouches.

predators
Animals that hunt other animals for food.

scavengers
Animals that eat food that is already dead.

To Learn More

BOOKS

Kenah, Katharine. *Super Marsupials: Kangaroos, Koalas, Wombats, and More*. New York: HarperCollins, 2019.

Schuh, Mari. *Opossums*. Minneapolis: Jump!, 2016.

NOTE TO EDUCATORS

Visit **www.focusreaders.com** to find lesson plans, activities, links, and other resources related to this title.

Index

B
babies, 20–21

D
dens, 19–20

E
eating, 13–14

F
fur, 9

P
predators, 16

T
tails, 5, 9–10

Answer Key: 1. Answers will vary; 2. Answers will vary; 3. C; 4. C